Bibliographic information published by the German National Library:

The German National Library lists this publication in the National Bibliography; detailed bibliographic data are available on the Internet at http://dnb.dnb.de .

Imprint:

Copyright © 2016 GRIN Verlag, Open Publishing GmbH
Print and binding: Books on Demand GmbH, Norderstedt Germany
ISBN: 9783668549593

This book at GRIN:

http://www.grin.com/en/e-book/377244/an-introductory-chemical-engineering-course-based-on-analogies-and-research-based

Zin Eddine Dadach

An Introductory Chemical Engineering Course Based on Analogies And Research-Based Learning

A Course Designed for Freshmen with Weak Science Background

GRIN Publishing

GRIN - Your knowledge has value

Since its foundation in 1998, GRIN has specialized in publishing academic texts by students, college teachers and other academics as e-book and printed book. The website www.grin.com is an ideal platform for presenting term papers, final papers, scientific essays, dissertations and specialist books.

Visit us on the internet:

http://www.grin.com/

http://www.facebook.com/grincom

http://www.twitter.com/grin_com

AN INTRODUCTORY CHEMICAL ENGINEERING COURSE BASED ON ANALOGIES AND RESEARCH-BASED LEARNING

Abstract

The main goal of this paper is to show that the use of analogies can be a very helpful tool in order to build a strong engineering foundation for freshmen who lack high school scientific background. To help students shift from the imaginary of the five analogies utilized in the course to the real pictures of some engineering concepts, the similarity to the analogy is followed by a simple lab experiment or a class activity. The final chapter of this course is related to energy efficiency to explain to freshmen who lack scientific background that their attitude could make them more efficient and lead them to success.

Keywords: Introduction Course, Chemical Engineering, Analogies, Research-Based Learning Activities.

1. Background

Freshmen who lack high school scientific background may not have the ability to link the theories learned during different chemical engineering courses and how to connect them to more general concept that can be applied to a wide variety of natural phenomena. Nowadays, this situation is further complicated since more sophisticated tools and formats are used to solve chemical engineering problems. For example, many students can memorize the ways to use difficult software without understanding the fundamental principles utilized by the software developers, to write the corresponding subroutines [1]. To overcome this situation, there is a need to adopt a deep approach to learning by trying routinely to relate course material to known situations. In this perspective, analogies could be very helpful to build conceptual bridges between what is familiar (an analog concept) and what is new (a target concept) [2]. The use of an analog concept could therefore help freshmen who lack some scientific background, develop images in their minds in order to visualize the physical phenomena behind each theory.

Many science teachers use analogies as a tool for more effective teaching. For example, Iveson [3] explained why counter-current is more efficient than co-current using a simple example of two basins used to clean dishes in stages. In a chemical engineering course, Fernandez-Torres [4] used different analogies (Number of male and female students in university, a drink with orange and water) to explain mass and molecular fractions to freshmen. In a chemistry course, Foos [5] used cartoon characters to explain electronegativity, redox reaction and electron delocalization. According to Yelamarthi et al. [6], increased student motivation, better participation in class and laboratory exercises, better rapport between the student and instructional group, increased creative thinking of the students and active student participation in providing valuable course feedback are some of the immediate positive outcomes in using analogies. However, it should be noted that analogies are useful only if the students are familiar with the analogy that instructors use and have the same understanding of the point that the instructors wish to convey. Thus, using the proper analogies and repetitively specifying the similarities and differences between the analog and target concepts is necessary [7].

Since engineering students need to work with real process applications, charts, diagrams, hands-on practices, and demonstrations concurrently with theory, equations, and words, they are encouraged to become active rather than passive learners [8, 9]. As a consequence, inquiry and research-based learning are fundamental for engineering education. With this in mind, lab experiments and class activities are the other tools that could help freshmen who lack scientific background to experience and "touch" the theories to be learned. For example, in an introduction to engineering course at Rowan University, a discussion on fluidized beds was associated to a laboratory experiment on fluidized bed coating developed at the university [10]. During a mechanical engineering course, a series of simple experiments on mechanics, fluid dynamics and electrical circuits has been developed to enhance freshmen's abilities to apply engineering concepts and to design experiments and to motivate the learning of computer tools [11]. According to Wankat and Orevitz [12], some benefits of inductive learning are motivation, problem identification, discovery, induction and the opportunity to build/test lab experiments that are memorable

In a previous paper, Dadach [13] used brain-body interactions as an analogy in a process control course in order to explain how controllers and control loops operate in chemical processes. Lab experiments and group activities were added as active learning strategies. The author used Ohm's law as a tool to quantify the impact of the learning strategy on the motivation of students. In this first published quantitative method, a "Dadach Motivation Factor" (DMF) was estimated for each student based on the Final Grade Point (FGP) and the Cumulative Grade Point Average (CGPA). The data showed that the use of analogies and the active learning strategy increased the performance of 69% of the students and enhanced the motivation of 40% of the students.

The objective of this second paper related to engineering education is to show the importance of using analogies and a research-based learning strategy in an introductory course for chemical engineering in order to help freshmen who lack some scientific background to transform the materials of the future courses of the department from opaque language into something they can visualize and integrate into their own knowledge bank.

2. Presentation of the course

This course serves as an introduction to concepts used in the analysis of chemical engineering problems. Based on analogies and research-based learning, it is intended to provide freshmen a global overview of the field of chemical engineering and help them visualize the elementary principles of Transport phenomena, Thermodynamics, Energy conservation and Energy efficiency. The Learning outcomes (LO's) of the course are related to the following chapters: (1) Electricity, (2) Fluid Dynamics, (3) Heat Transfer, (4) Mass Transfer & Solubility, (5) Thermodynamics, (6) Energy efficiency.

For every chapter, the teaching strategy followed the three main steps: (a) a basic introduction of the theory, (b) Similitude with the selected Analogy, (d) Research-based learning: lab experiments or class activities and (e) A conclusion. The lab experiments of the learning outcomes (1) to (4) include the following procedures:
1. Handout to read (10 minutes): Students are given a handout containing the main objective, the description of the apparatus, safety and procedure and a brief explanation of the theory.
2. Short lecture (20 minutes): The handout is explained and questions are answered.
3. First activity (Research-based Learning: 30-40 minutes): A team of three to four students collect the data while conducting the experiments and filled in the corresponding tables in the handout.
4. Second activity (Research-tutored Learning: 40 minutes): Every team analyses the collected data and investigates the relationship between the different variables of the system.
5. At the end of the activity, every team submits their handouts including calculations, answers to selected questions, discussions of results and conclusions.

3. Transport phenomena
3.1   General equation

Electrical charges, Momentum, Mass and heat transport all share a very similar framework where movement is motivated by a driving force within the system and is inversely proportional to the resistance located between two poles. As a consequence, these processes could be described using the following generalized relationship:

$$Flow = \frac{Driving\ Force}{Resistance}$$ (1)

3.2   Selected analogy

To make transport phenomena easier to understand, traffic situations on different roads of Abu Dhabi (Figure 1) are selected as the analogy to visualize movements during chemical engineering processes.

[picture removed for publication]

Figure 1: Traffic in a highway in Abu Dhabi [14]

3.3   Class activity

Some pictures showing cars on highways and streets of the city during different times of the day were distributed to the students. The objective of the activity was to analyze the following situations:

3.3.1 Situation #1: Before 7am, the corresponding pictures showed roads and highways with only a few cars. It was deduced that only few people had the motivation to go to work or study. Consequently, the weak "total driving force" could explain the small number of cars on the streets and highways. Based on equation (1), it could be concluded that the small "flow" of cars was not caused by the width of the road but rather by the weak "total driving force".

3.3.2 Situation #2: Around 7:30 am, most people have started to leave home and the corresponding pictures show that the number of cars suddenly increases. During this

4

rush hour, the total motivation was very high and it is easy for the students to link the larger number of cars on the highways to a higher "total driving force".

3.3.3. Situation #3: Pictures related to the traffic in the city also show that the streets are full of cars but that their number is much lower compared to the number of cars on the highways (previous pictures). As a consequence, for the same "total driving force", the width of the road made a difference on the number of cars. Analyzing equation (1), students link the small number of cars to the width of the road.

3.3.4 Situation #4: Finally, students have to imagine the situation around midnight where the total motivation is now close to zero. Streets and highways become very quiet. Based on equation (1), with no "driving force", any system becomes static.

From this analogy, the conclusion is that the number of cars increased by increasing the "total motivation "and decreased by decreasing the width of the roads. For the analogy, Equation (1) is rewritten as:

$$Flow\ of\ cars\ \propto \frac{Total\ motivation}{(1/Width\ of\ road)} \quad (2)$$

At the end of the activity, an explanation gets given to the students that people have individual motivation (individual driving force). The "Total motivation" is simply the sum of the "individual motivation". On the other hand, matter and energy in any system are under the same "total driving force".

3.4   Learning Outcome #1: Electricity and Ohm's law
3.4.1 Fundamentals
Voltage (V), current (I), and resistance (R) were first defined as shown in Figure 2.

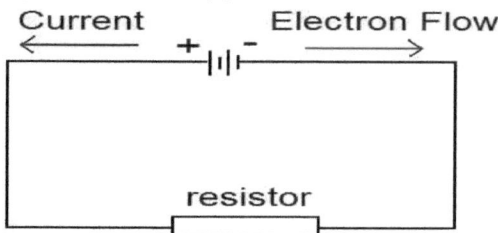

Figure 2: Flow of electrons [15]

Based on Figure 2: (1) Movement of electrons (I) in an electrical circuit looks like the flow of cars on the roads where electrons could be considered as "nano-cars", (2) the value of

5

the electrical resistance (R) could be linked to the width of the road, (3) the difference in voltage (V) represents the "total motivation" of people to take the road.

3.4.2 Research-based Learning

In order to shift from the pictures of the analogy to the basics of Ohm's law, a lab experiment was designed (Figure 3).

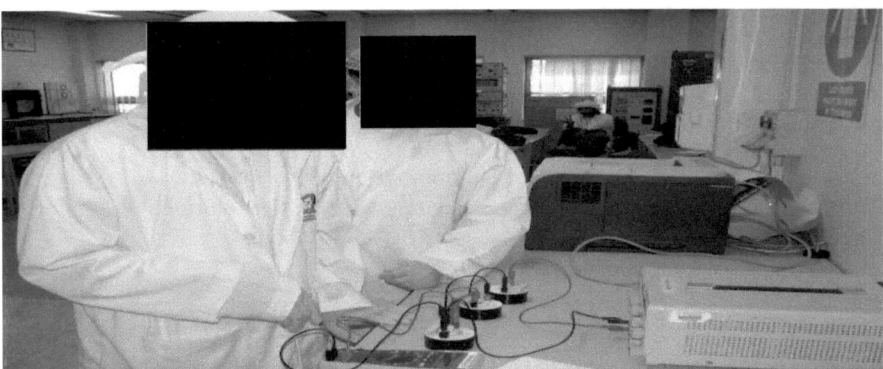

Figure 3: Lab experiment related to Ohm's law

The main objective of the lab experiment was to identify the flow, the driving force and the resistance of equation (1) in an electrical circuit. Students conducted the lab experiments using the following procedure:

1. Similar to situations #1 and #2 of the analogy, the current through the circuit was measured using a small electrical resistance of 5 Ω and selecting two different values of voltage (small and large). The experimental data are shown in Table 1

Table 1: Effects of increasing the voltage on the current

Analogy	Resistance (Ω)	Voltage (V)	Current (A)
Situation #1	5	12	2.4
Situation #2	5	80	16

2. Similar to situation #2 and #3 of the analogy, a large current source of 80V was selected and the intensity was measured for two electrical resistances of the circuit (Table 2).
3. Similar to situation #4 of the analogy, a zero voltage was utilized and the students noticed that there was no current even with a small electrical resistance. The system becomes static (Table 2).

Table 2: Effects of increasing the electrical resistance of the circuit on the current

Analogy	Voltage (V)	Resistance (Ω)	Current (A)
Situation # 2	80	5	16
Situation #3	80	100	0.8
Situation #4	0	5	0

3.4.3 Research-tutored Learning

During the second part of the activity, every group of students analyzed the collected data. In the first case, the intensity had increased by increasing the voltage and in the second situation, the intensity decreased when the resistance of the circuit increased. In concordance with the analogy, it was concluded that the flow of electrons is motivated by a potential in voltage (total motivation) and controlled by varying the value of the electrical resistance (width of the road) of the system. Based on Equation (1), the corresponding equation (2) was derived.

$$Flow\ of\ electrons \propto \frac{Voltage}{Electrical\ Resistance} \qquad (3)$$

Finally, equation (2) was linked to Ohm's law

$$I = \frac{Q}{t} = \frac{\Delta U}{R} \qquad (4)$$

Where: I= Electrical current, Q= electrical charges in Coulombs, t= time in seconds, R= Electrical resistance in Ohms, ΔU= Potential energy of a battery in Volts

3.5 Learning Outcome #2: Fluid Dynamics.

3.5.1 Fundamentals

After defining fluid flow, velocity, pressure and friction, the effects of pressure and friction on the movement of a fluid in a pipe were explained.

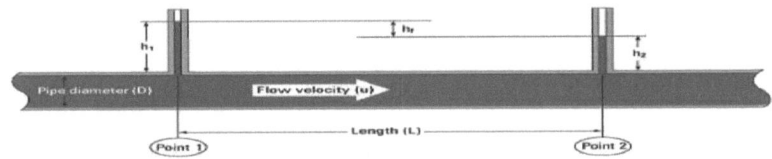

Figure 4: Friction in pipes [16]

Using Figure 4, the learning outcome #2 was also linked to the analogy: (1) The difference in pressure is considered as the "total motivation", (2) The movement of molecules (flow) is related to the flow of cars, (3) The effects of friction in pipes correspond to the width of the road.

3.5.2 Research-based Learning

As shown in Figure 5, the corresponding lab consist of two small bottles of propane usually used for a picnic. One bottle has high pressure and the other has low pressure. For the investigation, the two bottles are connected one after the other to a burner with a manual valve used to create friction in the system. The height of the flame is measured as an indication of the flow.

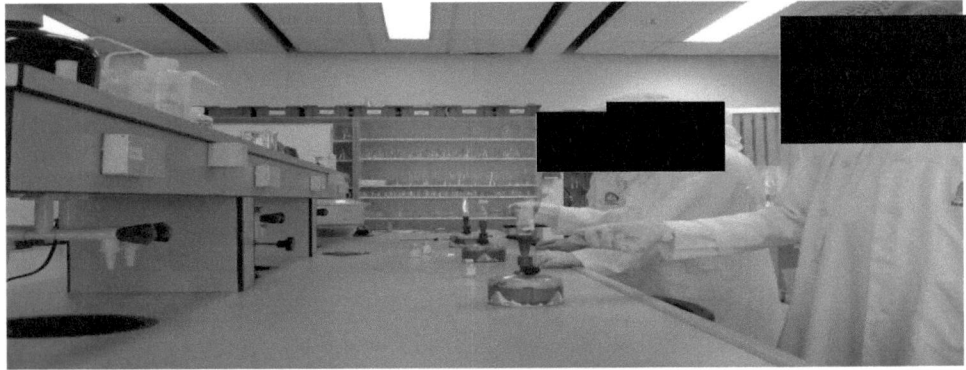

Figure 5: Lab experiment related to the flow of gas in a pipe

After the explanation of the hand-out, students conduct the lab experiment based on the following investigation:

8

1. During the first set of experiments (situations #1 and # 2 of the analogy), using a full opening of the valve, the bottle having low pressure and high pressure are utilized simultaneously and the height of the flame is measured (Table 3)

Table 3: Results of the first set of experiments

Analogy	Opening of the valve (%)	Pressure in bottle (psig)	Height of flame (cm)
Situation #1	100	10	15
Situation #2	100	50	➢ 50

2. The second set of experiments (situations #2 and #3 of the analogy) consist of using the bottle having high pressure and measuring the height of the flame for two positions of the valve (Table 4).
3. During the last experiment (Situation #4 of the analogy), students use an empty bottle and realize that, without a variation in pressure between the bottle and the atmosphere, there is no flame even with the valve fully open (minimum friction). The system becomes static (Table 4).

Table 4: Results of the second set of experiments

Analogy	Pressure in bottle (psig)	Opening of the valve (%)	Height of flame (cm)
Situation #2	50	100	➢ 50
Situation #3	50	5	7
Situation #4	0	100	0

3.5.3 Research-tutored Learning
During the second part of the activity, students analyze the collected data and realize that the flame is higher when the bottle is full (high pressure) and the height of the flame increases with the opening of the valve (decreasing friction). In agreement with the analogy, it is concluded that the flow of propane increases with increasing pressure in the

bottle (driving force) and decreases with the closing of the valve (increasing friction). After this picnic experiment, the students rewrite the general equation (1):

$$Gas\ Flow \propto \frac{P_B - P_{ATM}}{(\frac{1}{OV})} \qquad (5)$$

Where: P_B= Absolute pressure in bottle, P_{ATM} = Atmospheric pressure, OV= Percentage of the opening of the valve. At the end of the activity, the following equation (4), derived from Bernoulli's equation, was explained:

$$Flow = \sqrt{\Delta P}.\sqrt{\frac{2}{\rho}} x \sqrt{\frac{A_1^2 \times A_2^2}{(A_1^2 - A_2^2)}}. \qquad (6)$$

Where:ΔP= Differential pressure between the bottle and the atmosphere, A_1 and A_2 = Area of the pipe and valve respectively, ρ= Density of fluid

3.6 Learning Outcome #3: Heat Transfer
3.6.1 Fundamentals
After an introduction of the different types of heat transfer, it is explained that heat transfer by conduction depends on the thermal conductivity of metals and the difference in temperature between the two ends of the metal.

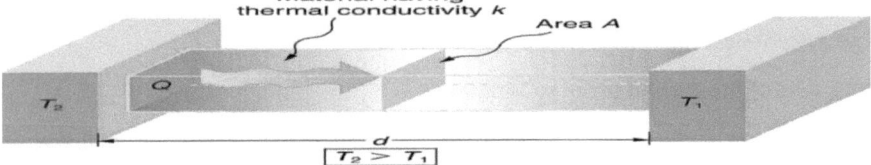

Figure 6: Heat transfer by conduction [17]

To link heat transfer by conduction (Figure 6) to the analogy, (1) the difference in temperatures represents the "total motivation" for heat flow, (2) the movement of heat waves (heat flow) is the flow of cars, (3) the value of the thermal conductivity corresponds to the width of the road.

3.6.2 Research-based Learning
As shown in Figure 7, the lab set-up consists of copper and steel rods, a beaker containing 100 grams of warm water and a beaker containing 100 grams of water at 0^0C.

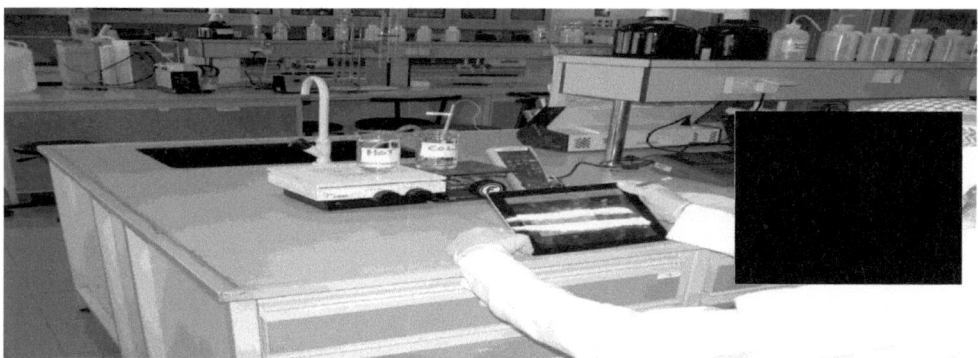

Figure 7: Experiment on heat transfer by conduction

Every group conduct the experiments following the investigation strategy:

1. Similar to situations #1 and #2 of the analogy, during the first set of experiments, copper is used and the temperature of the cold water is measured for two the different temperatures of warm water (20^0C and 100^0C).

Table 5: Experiments using copper for heat transfer by conduction

Analogy	Metal	Temperature warm water (^{0}C)	Final Temperature cold water (^{0}C)
Situation #1	Copper	20	3.7
Situation #2	Copper	100	16

2. Similar to situations #2 and #3 of the analogy, the temperature of the warm water of 100^0C is used and the temperature of the cold water is measured (Table 6) for the coper and steel simultaneously (Table 6).

3. Similar to situation #4 of the analogy, water in the two beakers has the same temperature of 0^0C and the students realize that the temperature stayed the same in the two beakers using copper (maximum conductivity). The system becomes static (Table 6).

Table 6: Experiments using copper and steel rod for heat transfer by conduction

Analogy	Metal	Temperature warm water (^{0}C)	Final Temperature cold water (^{0}C)
Situation #2	Copper	100	16
Situation #3	Steel rod	100	1.8
Situation #4	Copper	0	0

3.6.3 Research-tutored Learning

Analyzing the collected data during the second part of the activity, students find out that the temperature of the cold water reaches a maximum of 16°C when temperature of the warm water was 100°C and copper are utilized. On the other hand, the temperature of the cold water reaches a minimum of 1.8 °C when the temperature of the warm water is 100°C and steel rod is used. In order to create images that could help students understand the concept of conductivity, copper which has a thermal conductivity of 385 W/m.K, was compared to a car in a highway moving with high speed and steel, which has a thermal conductivity of 54 W/m.K, was compared to a car in a street moving with a much lower speed. As a consequence, the resistance to heat transfer by conduction is the inverse of the thermal conductivity. Based on equation (1), the following equation is derived by the students to explain their experimental data:

$$Heat\ by\ conduction = \frac{\Delta T}{1/Thermal\ conductivity} \tag{7}$$

Fourier's law of heat conduction (Thermal Ohm's law) is introduced and explained:

$$Q = \frac{\Delta T}{t/kA} \tag{8}$$

Where: ΔT= Difference in Temperature, t= Time, A= Surface area of the metal, k= Heat conductivity of the metal

3.7 Learning Outcome #4: Mass transfer & Solubility

3.7.1 Mass Transfer

Similar to heat transfer, mass transfer of a solute (A) in a solvent occurs from high concentration to low concentration. Using the convective mass transfer coefficient (k), the molecular flux (N_A) is represented by the equation:

$$N_A = k \ (C_{A1} - C_{a2}) \tag{9}$$

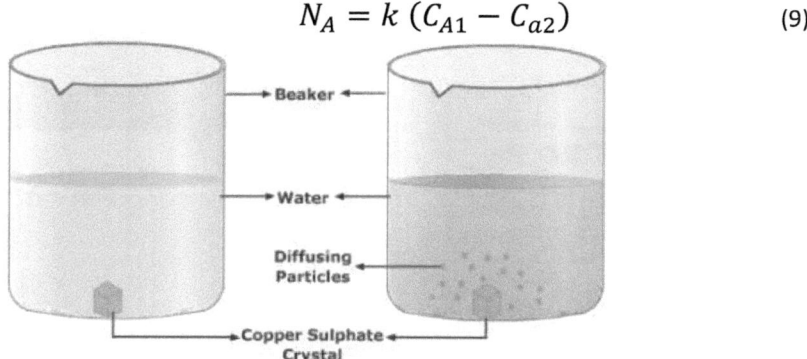

Figure 8: Driving force in mass transfer [18]

To link mass transfer (Figure 8) to the pictures of the analogy, (1) the difference in concentration corresponds to the "total motivation", (2) movement of molecules of solute is the flow of cars and (3) the value of the convective mass transfer coefficient (k) represents the width of the road.

3.7.2 Solubility

By definition, solubility is the capacity for one product (solute) to form a homogenous mixture (a solution) with another product (solvent). The objective of this learning outcome is to explain that solubility is related to molecular interactions. A second analogy is used in the course to create a bridge connecting the chemical affinity between different species and the affinity between people. The selected analogy for affinity is language. For example, in a foreign country, people who speak the same language stay together because they feel comfortable and can communicate with each other. It is then explained that, like people, two different chemical species can mix because of some affinities (like dissolves like).

3.7.3 Research-based Learning
The practical part related to mass transfer and solubility consisted of a lab experiment about liquid-liquid extraction (Figure 9).

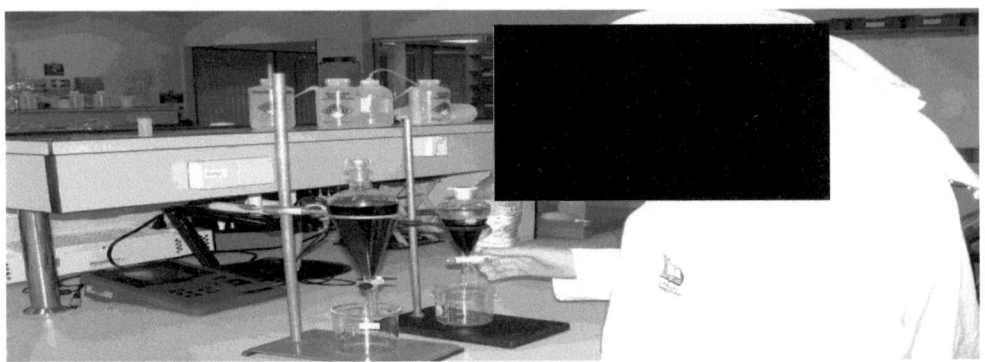

Figure 9: Experiment on Liquid-Liquid extraction

During the experiments, students first mix water with iodine to produce a purple colored solution. Cyclohexane is then added to the initial mixture. Students take note about the change in color of the two solutions after shaking the glassware several times during the experiment: (1) it is noticed that, after each mixing, the color of the cyclohexane becomes darker while the color of water becomes lighter. It is easier to realize that transfer of molecules of Iodine (dark purple color) took place from water to cyclohexane. (2) at the end of the experiment, water with a slight purple color is at the bottom of the separation funnel while cyclohexane with the deep purple color of iodine is at the top of the glassware.

3.7.4 Research-tutored Learning
During the second part of the activity, students use solubility tables in order to interpret the outputs of the experiment. The solubility of iodine in water at 20^0C is 0.3 g/100g and the solubility of iodine in cyclohexane is 2.8 g/100g. It is then concluded that iodine (non-polar) is slightly soluble in water (polar) because it has a weak interaction with the hydrogen bonded water molecules. However, because, cyclohexane (non-polar) has weaker solvent/solvent interactions than water, iodine found it easier to disrupt these interactions and insert itself between the molecules of cyclohexane.

14

3.8 Summary

At the end of the learning outcomes related to transport phenomena, a review of all the processes is covered during a class activity. Every group of students has to fill the blanks related to "Flow" "Driving Force" and "Resistance" of equation (1) for all the transport phenomena in the given Table 7.

Table 7: Review of Transport Phenomena

	Flow	Driving Force	Resistance
Analogy	Cars	Sum of individual motivation of people	1/ width of the road
Electricity	Electrons	Difference in voltage	Electrical resistance
Fluid dynamics	Molecules of propane	Difference in pressure	Friction
Heat transfer	Heat waves	Difference in temperature	1/ thermal conductivity
Mass transfer	Molecules of Iodine	Difference in concentration	1/(convective mass transfer coefficient)

4 Learning Outcome #5: Thermodynamics

Thermodynamics is the second topic in this introduction course. The First Law is about energy conservation and the second Law is about the natural direction of flows.

4.1 1st Law of thermodynamics

4.1.2 Fundamentals

The first law of thermodynamics is related to energy conservation, adapted for thermodynamic systems. The law of conservation of energy states that the total energy of an isolated system is constant; energy can be transformed from one form to another, but cannot be created or destroyed. To make energy conservation easier to understand, money in different currencies is selected as the third analogy.

4.1.3 Class activity

A class activity is organized where students have money in different currencies (UAE Dirham, European Euro and American dollar). In the beginning of the activity, students have only UAE dirhams and start to convert a certain amount of their money into

European currency and another amount into American dollars using the real rate. At the end of the activity, they convert all their money back to UAE Dirham and the following equation is derived:

Money (Dirham) + Money (€) + Money ($) = Total amount = Constant   (10)

Students also realize that, in any money conversion process, the amount of money of one currency decreases while the amount of money of the other currency increases. During the second part of the activity, the principle of mechanical energy conservation (Bernoulli's equation) is introduced and students are asked to solve the following problem: "Water flows through a horizontal pipe. At one end, the water in the pipe has a pressure of 150,000 Pascal (Pa), a speed of 5.0 m/s. At the other end, the speed of the water is 10 m/s. The pressure at the second end should be calculated using equation (8)

$$p\ (Pressure) + \frac{1}{2}.\rho.v^2(\ kinetic) + \rho.g.h\ (\ potential) = Constant \quad (11)$$

Where p= Pressure, ρ= Density, v= velocity, g= gravity, h= height. Similar to the analogy, the kinetic energy increases by increasing the velocity from 5 m/s to 10m/s. As a consequence, the pressure will decrease from 150,000 Pa to 112,500 Pa.

4.1.5 Conclusion
Like the total amount of money, total energy is then conserved in any system but can be converted from one form to another. It is also explained that, like the loss of some money during money transfer in a different currency, some amount of energy is also lost during any energy conversion process mainly by friction.

4.2 Second Law of thermodynamics

4.2.1 Fundamentals
The second law of thermodynamics is also discussed in this introduction course. The second law tells us that energy naturally flows from a higher concentration to a lower concentration. In order to explain that natural movements are only from high energies to low energies, the fourth analogy used in this course is physical strength.

4.2.2 Class activity
During the first part of the class activity, a strong student and a weak student (as a reference) of each group are asked to perform arm-wrestling. As expected, the strong student displaces the arm of the weak student (reference). A much stronger student is

then invited to conduct the same arm wrestling activity with the strong student. Compared to the third student, the strong student is now weak and could not displace the arm of the other student. It is clear that energy is just a potential and work can be performed only if the other part has a lower energy. Following the analogy, a second activity is organized where students are asked to study a steam table. Enthalpy is the selected energy in this activity (Figure 10).

Saturated Steam								
Pressure in Absolute PSI								
Abs press (psia) lb per in²	Temp °F	Specific volume ft³ / lbm		Enthalpy btu / lbm		Entropy btu / lbm × °F		Abs press (psia) lb per in²
		Water V_f	Steam V_g	Water h_f	Steam h_g	Water S_f	Steam S_g	
0.08865	32.018	0.016022	3302.4000	0.0003	1075.5	0.0000	2.1872	0.08865
0.250	59.323	0.016032	1235.5000	27.382	1067.4	0.0542	2.0967	0.250
0.500	79.586	0.016071	641.5000	47.623	1096.3	0.0925	2.0370	0.500
1.000	101.74	0.016136	333.6000	69.730	1105.8	0.1326	1.9781	1.000
3.000	141.47	0.016300	118.7300	109.420	1122.6	0.2009	1.8864	3.000
6.000	170.05	0.016451	61.9840	138.030	1134.2	0.2474	1.8294	6.000
10.000	193.21	0.016592	38.4200	161.260	1143.3	0.2836	1.7879	10.000
14.696	212.00	0.016719	26.7990	180.170	1150.5	0.3121	1.7568	14.696
15.000	213.03	0.016726	26.2900	181.210	1150.9	0.3137	1.7552	15.000
20.000	227.96	0.016834	20.0870	196.270	1156.3	0.3358	1.7320	20.000
25.000	240.07	0.016927	16.3010	208.520	1160.6	0.3535	1.7141	25.000
30.000	250.34	0.017009	13.7440	218.900	1164.1	0.3682	1.6995	30.000
35.000	259.29	0.017083	11.8960	228.000	1167.1	0.3809	1.6872	35.000

Figure 10: Saturated steam Table [14]

It is found that under some conditions at the beginning of the tables, the enthalpy of the saturated liquid water is very close to zero. The values of the enthalpy became higher by increasing temperatures. It is explained that the pressure and temperature that give a zero value for the enthalpy are the "reference conditions" used in the table. They are then advised to always check the reference of the tables before using any value for calculation. To continue the class activity, the two following temperatures are selected (T1= 32.018^{0}F, P1= 0.08865 psia and T2= 213.03^{0}F, P2= 15 psia). The corresponding enthalpies of the saturated liquid at T1 and saturated steam at T2 are respectively h_{l1}= 0 btu/lbm and h_{g2}= 1150.9 btu/lbm. Based on this analogy, it is explained that production of electricity from a steam turbine could happen only from a steam at high energy (h_{g2}) to become a steam at lower energy (h_{l1}). Using the concept of energy conservation, the following equation is derived:

17

$$h_{g2} = h_{l1} + Work\ (turbine) \qquad (12)$$

Conclusion

Reviewing Learning outcome 1 to 4, movements are always from a higher concentration to a lower concentration". Finally, to explain that the field of chemical engineering is part of the natural sciences, the same concepts of transport phenomena and thermodynamics are utilized to describe the water cycle in nature and the transformation of light energy from the sun to chemical energy in the plants and also how people can perform work by consuming carbohydrates as the chemical energy of the plants.

5. Energy Efficiency

5.1 Fundamentals

In chemical plants, energy is used as work or heat by equipment like compressors, pumps, turbines, heat exchangers, boilers and condensers located at the interface between the fluid in the process and the utilities (energy sources) used by the process. For these equipment, energy efficiency could be defined as the ratio of total enthalpy change of the transported fluid (ΔH) over the energy (E) used:

$$\eta = \frac{\Delta H}{E} \qquad (13)$$

5.2 Class activity

To motivate students to take action for sustainable development, the last analogy used in this introduction course is the well-known story of the "Tortoise and the Hare". The hare moved faster but in different directions. On the other hand, the tortoise was moving slowly but steadily towards the final destination. Based on the story, a P-V diagram is used to draw the path of both the hare and the tortoise between the two fixed points (Figure 11) which are the beginning and the end of the imagined race.

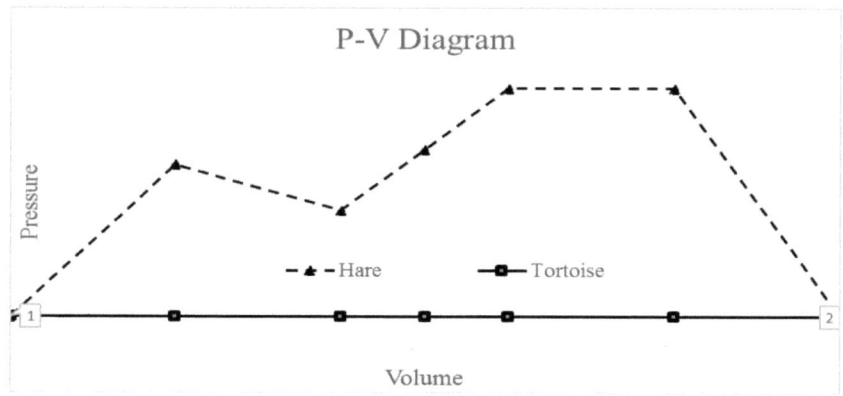

Figure 11: PV diagram showing the race between the tortoise and the hare

Analyzing the P-V diagram, students realize that the area under the path used by the tortoise is much smaller than the one utilized by the hare for the same change of states. Since the area under the state (1) and state (2) is the work PV used by both animals, the following energy efficiencies could be written:

$$\eta_{Tortoise} = \frac{\Delta H}{W_{Tortoise}} \tag{14}$$

$$\eta_{Hare} = \frac{\Delta H}{W_{Hare}} \tag{15}$$

It is concluded that the hare wasted too much energy. The tortoise is then more efficient and won the race even with a smaller capacity to run. My final point to the freshmen who lack some scientific background is that they should make efforts and keep learning because their attitude and motivation is more important

6. Conclusion

The objective of the use of analogies and a research-based teaching strategy in an introduction to chemical engineering is to give to the students a global picture of their field. My aim is to help students imagine real situations behind every mathematical equation and to properly use the theory. For those who did not like the "flavor" of the field, they can choose an alternative option. The others, who enjoy the "flavor" during the introductory course, will easily recognize the analogies and limitations in each theory and concentrate on learning the different applications thereof during future courses or "the meal".

Acknowledgments

The author would like to thank the Higher Colleges of Technology of UAE for offering this opportunity to go back and teach the basic concepts of Chemical Engineering. I should not forget to thank all my students who inspire me and teach me so much every day.

References

1. Falconer, J.L.; "Use of concepts and Instant Feedback in Thermodynamics", Chem. Eng Ed. 38(1), 64 (2004)

2. M. Kearney and K. Young, An emerging learning design based on analogical reasoning, Proceedings of the 2nd International LAMS Conference Practical Benefits of Learning Design, 2007: pp. 51-61.

3. S. Iveson, Explaining why Counter-current is more efficient than Co-current, Chem. Eng. Ed., 36(4), 2006, pp. 257- 263.

4. M.J. Fernandez- Torres, Those little tricks that help students to understand basics concepts in Chemical Engineering, Chem. Eng. Ed, 39 (4),2005, pp. 302-307.

5. C. W, J. Foos, Making Chemistry Fun to Learn, Literacy Information and Computer Education Journal (LICEJ), Volume 1, Issue 1, March 2010.

6. K. Yelamarthi, S. Ramachandran, P. R. Mawasha, and B. A. Rowley, The practical use of analogies to mentor the engineer of 2020, American Society for Engineering Education, March 31-April 1, 2006.

7. S. M. Glynn, Making science concepts meaningful to students: Teaching with analogies. In S. Mikelskis-Seifert, U. Ringelband, & M. Brückmann (Eds.), Four decades of research in

science education: From curriculum development to quality improvement, Münster, Germany: Waxmann, 2008, pp. 113-125.

8. D. Fraser, The Phumelela Project: improving the success of engineering students. Aalborg: proceedings of the 36th SEFI Annual Conference, 2008, pp. 35

9. R. M. Felder and L.K. Silverman, Learning and Teaching Styles in Engineering Education," Eng. Ed., 78 (7), 1988 pp. 674-681

10. R. P. Hesketh, C. S. Slater, S. Farrell, and M. Carney, "Fluidized Bed Polymer Coating Experiment," Chem. Eng. Ed. 36(2) 138 (2002).

11. Jed S. Lyons and John S. Brader; Using the learning cycle to develop freshmen's abilities to design and conduct experiments; International Journal of Mechanical Engineering Education, April 2004; vol. 32, 2: pp. 126-134.

12. P. Wankat and F.S. Oreovicz, Teaching Engineering, New York, McGraw-Hill, 1993.

13. Z. Dadach , quantifying the effects of an active learning strategy on the motivation of students., International Journal of Engineering Education, Vol. 29, No. 4, pp. 904–913, 2013

14. Emirates News Staff, Speed cameras on Abu Dhabi-Dubai highway set at 140kmph (2011)

15. Ohm's law from Wikipedia

16. Pipes and Pipe Sizing; International site for Spirax Sarco

17. Matt Williams, The Science of Heat Transfer: What Is Conduction?; Universe Today (2014)

18. Diffusion from Wikipedia

19. ASME Steam Tables: Compact Edition (Crtd) 1st (first) by ASME Research and Technology Committee on Water and Steam in (2006).